Imprint:

Copyright © 2017 GRIN Verlag, Open Publishing GmbH
Print and binding: Books on Demand GmbH, Norderstedt Germany
ISBN: 9783668382640

This book at GRIN:

http://www.grin.com/en/e-book/351562/molecular-phylogenetic-studies-of-bitter-gourd-momordica-charantia-and

Prem Jose Vazhacharickal, Sajeshkumar N.K., Jiby John Mathew, Linta Jose, Lakshmi M. Nair

Molecular phylogenetic studies of bitter gourd (Momordica charantia) and ivy gourd (Coccinia grandis) and family comparison using rbcL sequence analysis

GRIN Publishing

Molecular phylogentic studies of bitter gourd (*Momordica charantia*) and ivy gourd (*Coccinia grandis*) and family comparison using rbcL sequence analysis

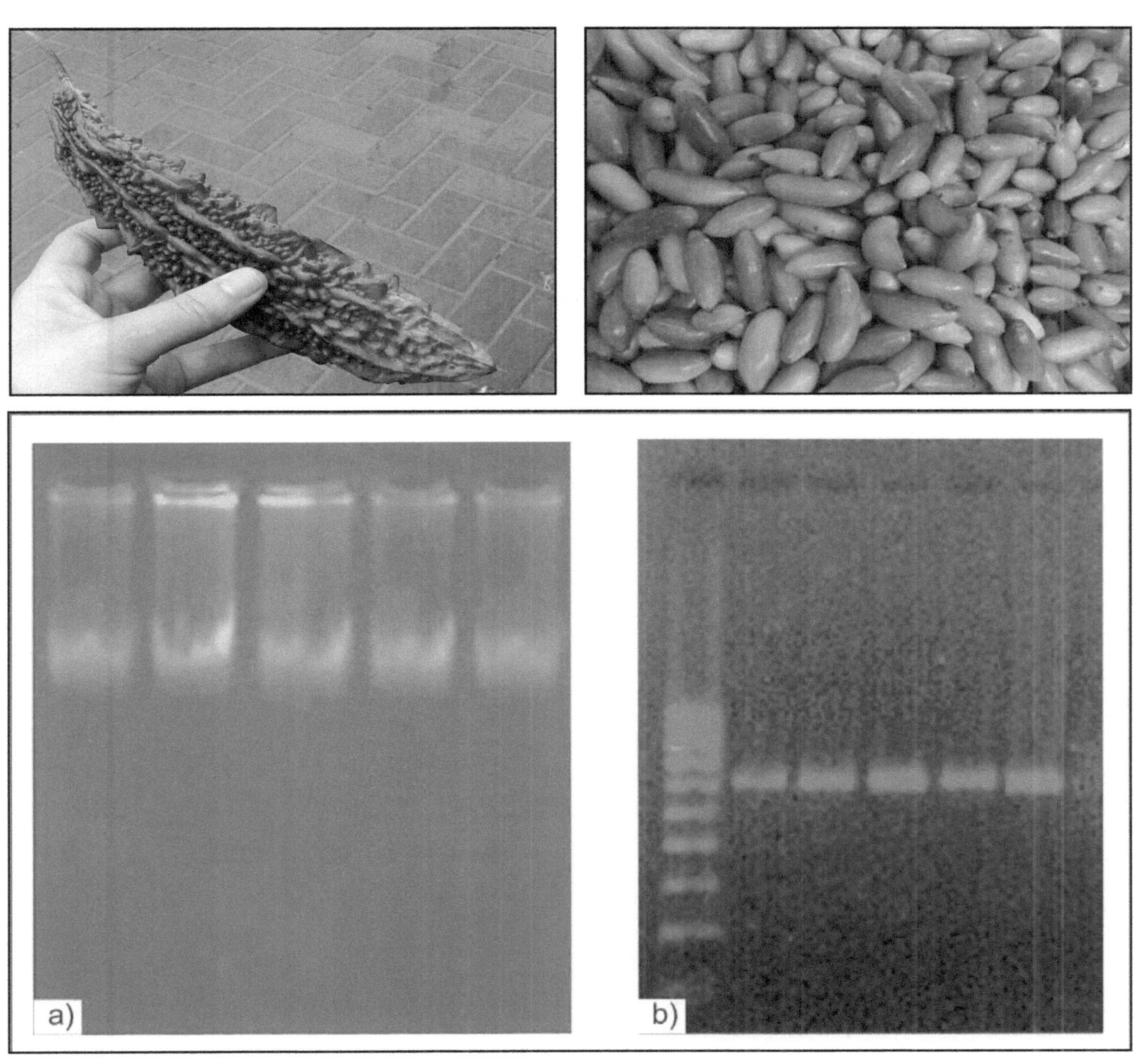

Prem Jose Vazhacharickal, Sajeshkumar N.K, Jiby John Mathew, Linta Jose and Lakshmi M Nair

Acknowledgements

Firstly we thank **God Almighty** whose blessing were always with us and helped us to complete this project work successfully.

We wish to thank our beloved Manager **Rev. Fr. Dr. George Njarakunnel,** Respected Principal **Dr. Joseph V.J,** Vice Principal **Fr. Joseph Allencheril,** Bursar **Shaji Augustine** and the Management for providing all the necessary facilities in carrying out the study. We express our sincere thanks to **Mr. Binoy A Mulanthra** (lab in charge, Department of Biotechnology) for the support. This research work will not be possible with the co-operation of many farmers.

We are gratefully indebted to our teachers, parents, siblings and friends who were there always for helping us in this project.

Prem Jose Vazhacharickal*, Sajeshkumar N.K, Jiby John Mathew, Linta Jose and lakshmi M Nair

*Address for correspondence
Assistant Professor
Department of Biotechnology
Mar Augusthinose College
Ramapuram-686576
Kerala, India
premjosev@gmail.com

Table of contents

Table of figures

List of abbreviations

BLAST	: Basic local alignment search tool
CTAB	: Cetyl trimethylammonium bromide
FASTA	: Fast All
HTML	: Hyper text markup language
$MgCl_2$	: Magnesium chloride
NJ	: Neighbour joining
PCR	: Polymerase chain reaction
RAPD	: Random amplified polymorphic DNA
rbcL	: Ribulose bisphosphate carboxylase large chain
Rubisco	: Ribulose-1, 5-bisphosphate carboxylase oxygenase
SPSS	: Statistical package for social sciences
Tag	: *Thermus aquaticus*

Molecular phylogenetic studies of bitter gourd (Momordica charantia) and ivy gourd (Coccinia grandis) and family comparison using rbcL sequence analysis

Prem Jose Vazhacharickal[1]*, Sajeshkumar N.K[1], Jiby John Mathew[1], Linta Jose[1] and Lakshmi M Nair[1]

[1]Department of Biotechnology, Mar Augusthinose College, Ramapuram, Kerala, India

* Corresponding author; premjosev@gmail.com

Abstract

The Cucurbitaceae family is the one of the economically important group of plants in the tropics and subtropics. Molecular phylogenetic analysis was advanced after the introduction of molecular markers which give much precise results in analysis. Our current study based on the amplification of RuBisco enzyme using rbcL primer and subsequent validation using BLAST, FASTA and CLUSTAL-W in pumpkin and winter melon. The isolated and purified DNA samples were PCR amplified using rbcL primer and later sequenced using ABI Prism 377 DNA sequencer. Multiple sequence alignment algorithms and distance matrix were constructed using rbcL sequences in FASTA format were retrieved from GenBank. Phylogenetic tree was created using the distance based neighbour joining (NJ) and clustering algorithms method. The agarose gel electrophoresis were used to separate the isolated as well as PCR amplified DNA samples. The sequences obtained after sequencing were subjected to BLAST similarity search and multiple sequence alignment using CLUSTAL-W. The construction of divergence matrix and phylogenetic dendrograms revealed two main groups, one consisting of MC and other Momordica species while the other group consisted of three subgroups of CM, CG and BH having members of Cucurbita, Coccinia and Momordica respectively. The reliability of molecular phylogenesis can be affected by myriad, long branch attraction, saturation and taxon sampling problems. So the selection of the models plays a key step in the success analysis. In general, the present study unambiguously confirmed the existence of *Coccinia grandis* as a distinct species and morphologic separation between *Coccinia grandis* and other selected members of Cucurbitaceae as separate species was evident. Among the species included in the present study *Coccinia grandis* undergone maximum divergence during the process of evolution. More detailed work on the evolution within the Cucurbitaceae needed for crop improvement.

Keywords: Clustering algorithms; rbcL sequences; Multiple sequence alignment.

1. Introduction

Crop improvement has been advanced a lot with the evolution of Genetics, genomic and plant breeding. The Cucurbitaceae family has 118 genera and 825 species mainly distributed in the tropics and subtropics (Esteras et al., 2011; Jeffrey, 2005) which were characterized by five angled stem and coiled tendrils (Kocyan et al., 2007). The genus Cucurbita (2n = 2x = 40) has a great diversity in physiological and morphological characteristics (Robinson and Decker-Walters, 1997; Goldman, 2004; Gong et al., 2013). In the gourd family (Cucurbitaceae), flowers are unisexual with both monecious and dioecious plants (Schaefer and Renner, 2010). Research on Cucerbitaceae usually based on tendril branching (Jeffrey, 2005), pollen structure and seed coat characterization (Kocyan et al., 2007). The Cucurbitaceae has been considered as an ideal candidate for genetic scrutiny using molecular marker tools (Lebeda et al., 2007; Gong et al., 2013).

Momordica consist of 59 species, most of them are perennial climber, with monecious and dioecious species (Palada and Chang, 2003; Lokesha and Vasudeva, 2001; Schaefer and Renner, 2010). Molecular marker based researches were usually geographic, country specific while some research were done across the globe for worldwide genetic variations (Gwanama et al., 2000; Paris et al., 2003; Wu et al., 2011; Gong et al., 2012). Phylogenetic studies based on allozyme (Wilson, 1989), ribosomal intergenic spacer probes (Ruiz and Hemleben, 1991), chloroplast DNA (Wilson et al., 1992), and RAPDs (Baranek et al., 2000; Ferriol et al., 2003) are reliable and efficient tools for evolutionary relationships. Nuclear DNA markers are more reliable than organelle DNA due to the inheritance from both parents as well as meiotic recombination (Gong et al., 2013). Rubisco (Ribulose-1, 5-bisphosphate carboxylase oxygenase) is an enzyme found in plant chloroplast, has a vital function in Calvin cycle and glucose synthesis.

Differences in DNA sequences are used to determine the evolutionary relationship which based on phylogenic tree is a major part in molecular systematic. Comparison of homologous sequences for genes using sequence alignment techniques and data base search is a good indicator of degree of divergence during evolution. Given lacking information about molecular phylogeny of bitter gourd and ivy gourd, our objectives were to (1) amplify the genes producing the RuBisco enzyme using rbcL primer, (2) validation of the amplified genes using similarity searches including BLAST and FASTA and (3) interpretation of the software algorithms using CLUSTAL-W.

2. Materials and methods

2.1 Plant samples collection

Fresh plant materials of bitter gourd (*Momordica charantia*) and ivy gourd (*Coccinia grandis*) was collected from a local farm in Melukavu (Figure 1). Fresh leaves samples were collected in pre-sterilized autoclavable polyethylene bags (Himedia Laboratories, Mumbai, India) and stored in insulated boxes with gel ice packs during transportation. The samples were stored at -20°C till further analysis. For the extraction of DNA, alkaline lysis method (Kidwell and Osborn, 1992) followed by modified CTAB (Porebski et al., 1997) method. The quality of the isolated DNA was determined spectrophomertically using a double beam UV spectrophotometer (118, Systronics India Ltd, Ahmedabad, India) as well as 8% agarose gels, 1x TBE buffer in an agarose electrophoresis chamber (Geni Pvt, Bangalore, India).

The amplification of the extracted DNA were done using PCR machine (MJ Mini, BioRad, Gurgaon, India) according to the protocol of Zang and Renner (2003) using Taq Polymesase

(Geni Pvt, Bangalore, India) and rbcL primer (Geni Pvt, Bangalore, India). The amplification performed in 25 µl of 25 µmol l^{-1} MgCl$_2$ solution, 2.5 µl 10 x buffer (Geni Pvt, Bangalore, India), 2 µl of a 2.5 µmol l^{-1} dNTP solution, 1 µl of rbcL primer at 10 pmol µl^{-1}, 1 unit (0.2 µl) of Taq Polymerase, and 1 µl of extracted DNA (Haas et al., 2003; Rychlik et al., 1990). The forward primer rbcL a F (5' ATGTCACCACAAACAGAGACTAAAGC 3') and reverse primer rbcL a R (5' GTAAATCAAGTCCACCACG 3').

After the PCR amplification, the PCR products were separated using agarose gel (1.5%) electrophoresis with a 100 base pair standard ladder DNA of 100 to 1000 BP (Geni Pvt, Bangalore, India). After agarose gel electrophoresis, the PCR products were trimmed from the gel, subsequently the gel was dissolved and later DNA was precipitated using 98% ethanol as precipitating agent (Meier et al., 1996; Erlich, 1989; Saiki, 1989). The purified DNA samples were later sequenced using ABI Prism 377 DNA sequencer (Applied Biosciences, NY, USA).

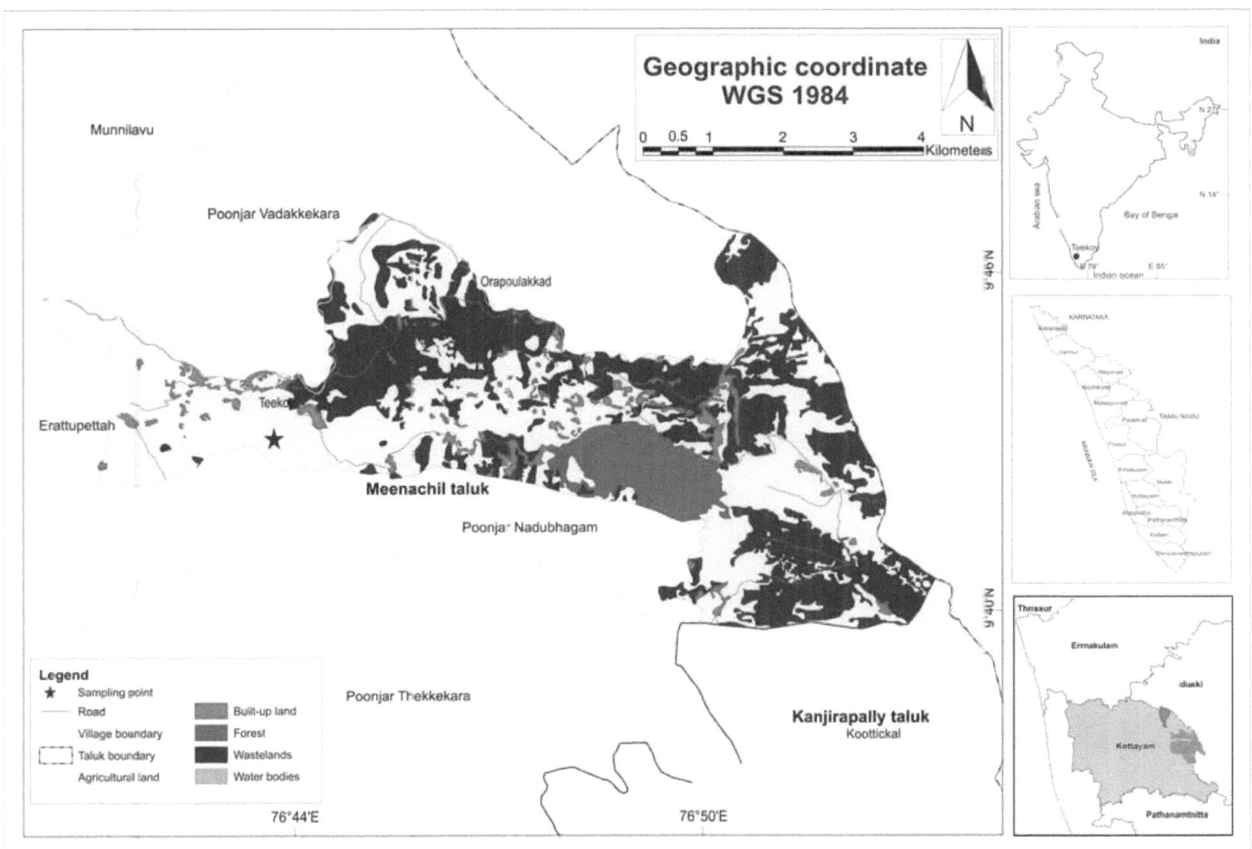

Figure 1. Map of the sample collection area represented by a star symbol.

Figure 2. a) Bitter gourd (*Momordica charantia*) fruits, b) Bitter gourd flowers, c) *Coccinia grandis* developing fruit, d) *Coccinia grandis* harvested mature fruits, e) winter melon (*Benincasa hispida*) fruit, f) *Cucurbita maxima* fruit. Photo courtesy: Wikipedia.

Table 1. Components of the PCR reaction mixture

Components	Volume/reaction
De-ionized water	16.5µl
Taq buffer (10X)	2.5 µl
Template DNA (20ng/µl)	2.0 µl
MgCl$_2$ (15Mm)	1.0 µl
dNTPs mix (10mMm each)	1.5 µl
Primer mix (10pm/µl)	1.0 µl
Forward primer	0.5µl
Reverse primer	0.5µl
Taq DNA polymerase (5U/µl)	0.5 µl
Final volume	25.0 µl

Table 2. PCR cycle for rbcL amplification

Cycle	Step	Process	Temperature	Time period
1	1	Initial denaturation	95°C	3 minutes
1	2	Denaturation	95°C	30 seconds
1	3	Annealing	55°C	30 seconds
1	4	Primer extension	72°C	30 seconds
2-32		Repeat steps 2 to 4 for 32 times		
33	5	Final elongation	72°C	10 minutes
		End or 4°C for ever		

2.2 Analysis of the sequences BLAST and Clustal-W

Basic Local Alignment Search Tool (BLAST) is used sensitive sequence similarity between sequences. The sample sequences in FASTA format were entered in field specified on BLAST algorithms home page. The BLAST output parameters were set to pre-determined settings and final output in Hyper Text Mark Up (HTML) format were used for further bioinformatics analysis.

Multiple sequence alignment programme Clustal-W 2.1 (University College of Dublin, Ireland). The process of Clustal multiple sequence analysis was advanced through pair wise alignment, construction of a distance matrix and display of base saturation in the matrix. The results were based on the pair wise analysis of 19 sequences using Maximum Composite Likelihood Method (Thompson et al., 1994; Chenna et al., 2003). Codon positions included were 1st + 2nd + 3rd + non coding sequences. Positions containing gaps and missing data were eliminated from the data set using complete deletion option to achieve high resolution phylogenetic tree construction.

In the case of bitter gourd, for precise results in the tree construction, 17 cucurbit samples and their respective rbcL sequences (GenBank) were included in the multiple sequence alignment algorithms along with *Momordica charantia*, *Cucurbita maxima*, *Coccinia grandis* and *Benincasa hispida*. For pumpkin, 15 cucurbit (other than *Cucurbita argyrosperma*) and an out-group Solanum melongena (brinjal) and their rbcL sequences in FASTA format were retrieved from GenBank. These were included in the multiple sequence alignment algorithms and in construction of distance matrix. The purpose of out-group is to increase the rate of interpretability of phylogenetic tree which shows maximum rate of evolutionary divergence.

Table 3. Sequences in FASTA format retrieved from GenBank for multiple sequence alignment algorithms and construction of distance matrix for winter melon.

No	Analysed sample	Retrieved sample species from GenBank
1	*Momordica charantia*	*Momordica foetida, Momordica cochinchinesis*
2	*Benincasa hispida*	*Benincasa fistulosus, Benincasa colocynthis*
3	*Cucurbita maxima*	*Cucurbita argyrosperma, Cucurbita palmate, Cucurbita foetidissima*
4	*Coccinia grandis*	*Coccinia quinqueloba, Coccinia adoensis, Coccinia mackenii*

Table 4. Sequences in FASTA format retrieved from GenBank for multiple sequence alignment algorithms and construction of distance matrix for pumpkin.

No	Analysed sample	Retrieved sample species from GenBank
1	*Momordica charantia*	*Momordica foetida, Momordica cochinchinesis*
2	*Benincasa hispida*	*Benincasa fistulosus, Benincasa colocynthis*
3	*Cucurbita maxima*	*Cucurbita argyrosperma, Cucurbita palmate, Cucurbita foetidissima*
4	*Coccinia grandis*	*Coccinia quinqueloba, Coccinia adoensis, Coccinia mackenii*
5		*Solanum melongena* (out-group)

From the preliminary results of multiple sequence alignment and distance matrix, non-aligned and mismatched sequences were removed for better reliability. Further this makes the understand ability of evolutionary convergence and divergences of the selected samples as well as with the other 11 common cucurbits included.

2.3 Construction of phylogenetic tree

The evolutionary relationships can be best studied using a phylogenetic tree and was created using the distance based neighbour joining (NJ) and clustering algorithms. Multifurcating phylogenetic tree was constructed based on creating a distance matrix starting from the alignment and pair wise distances calculated between sequences from the obtained matrix.

2.4 Statistical analysis

Descriptive statistics using SPSS 12.0 (SPSS Inc., Chicago, IL, USA) were conducted to summarize the data and graphs were generated using Sigma Plot 7 (Systat Software Inc., Chicago, IL, USA).

3. Results

The isolated DNA were checked spectrophotometrically and with a purity values of 1.67 and with a concentration of 4,395 μgml^{-1}. The agarose gel electrophoresis were used to separate the isolated as well as PCR amplified DNA samples (Figure 2). The sequences obtained after sequencing were subjected to BLAST similarity search and multiple sequence alignment using CLUSTAL-W.

The construction of divergence matrix and phylogenetic dendrograms revealed two main groups, one consisting of MC and other Momordica species while the other group consisted of three subgroups of CM, CG and BH having members of Cucurbita, Coccinia and Momordica respectively.

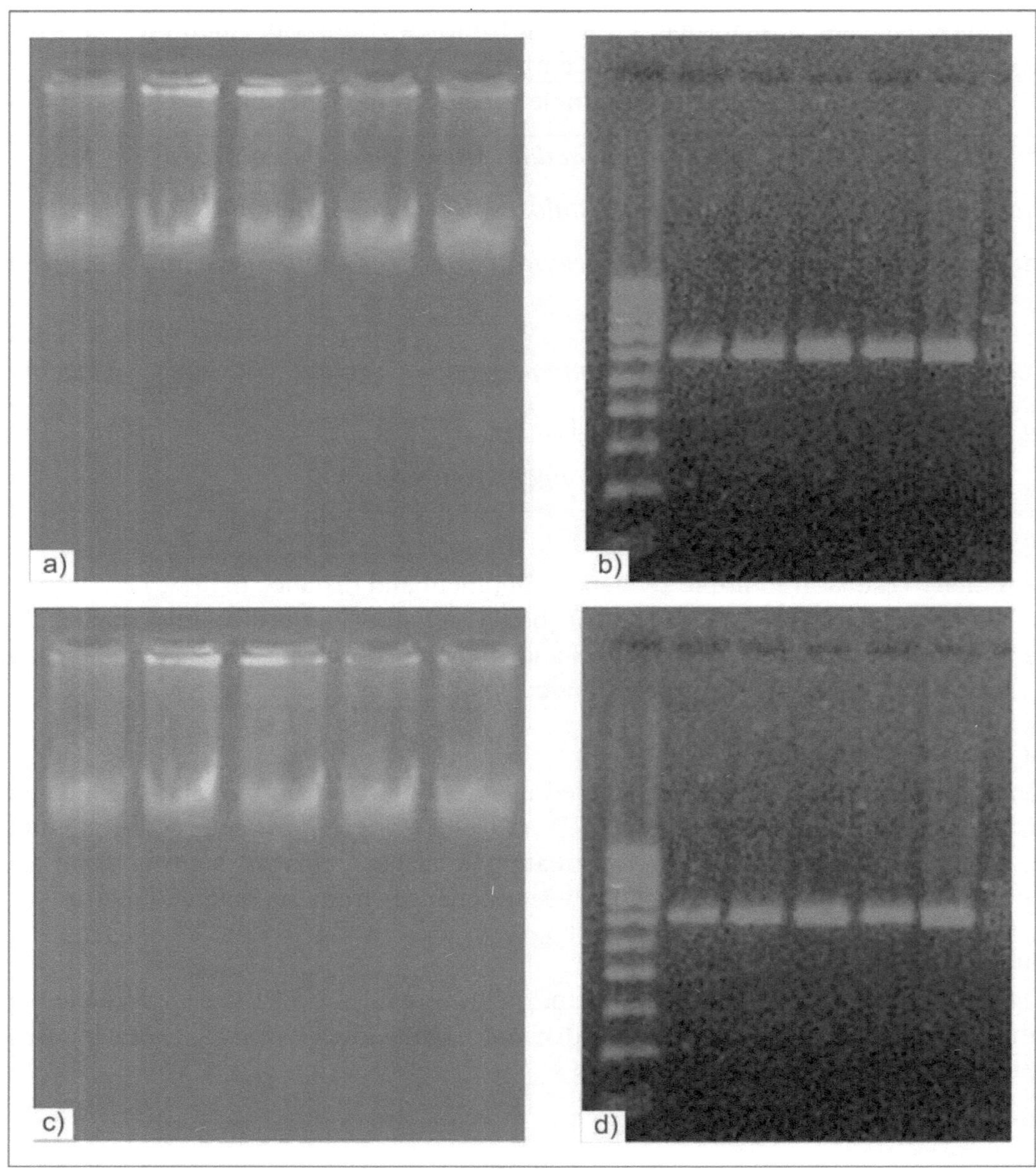

Figure 3. a) Agarose gel electrophoresis of the isolated DNA samples and b) PCR amplified DNA samples from bitter gourd. c) Agarose gel electrophoresis of the isolated DNA samples and d) PCR amplified DNA samples from ivy gourd.

Accession	Description	Max score	Total score	Query coverage	E value	Max ident
DQ535760.1	Momordica charantia ribulose-1,5-bisphosphate carboxylase/oxygenase la	830	830	100%	0 0	100%
DQ535829.1	Momordica foetida voucher H. Schmidt et al. 1978 (MO) ribulose-1,5-bisp	813	813	100%	0 0	99%
DQ535740.1	Baijiania borneensis ribulose-1,5-bisphosphate carboxylase/oxygenase lar	797	797	100%	0 0	99%
DQ501258.1	Baijiania yunnanensis ribulose-1,5-bisphosphate carboxylase/oxygenase l	797	797	100%	0 0	99%
JN407272.1	Momordica charantia isolate shawpc0969L ribulose-1 5-bisphosphate carl	795	795	95%	0 0	100%
JF944712.1	Thladiantha villosula voucher LiHT10173 ribulose-1,5-bisphosphate carbo	791	791	100%	0 0	98%
JF944711.1	Thladiantha villosula voucher LiHT10179 ribulose-1,5-bisphosphate carbo	791	791	100%	0 0	98%
JF944710.1	Thladiantha villosula voucher LiHT10180 ribulose-1,5-bisphosphate carbo	791	791	100%	0 0	98%
JF944709.1	Thladiantha villosula voucher LiHT10181 ribulose-1,5-bisphosphate carbo	791	791	100%	0 0	98%
JF944708.1	Thladiantha villosula voucher LiHT203 ribulose-1,5-bisphosphate carboxyl	791	791	100%	0 0	98%
JF944707.1	Thladiantha villosula voucher LiHT214 ribulose-1,5-bisphosphate carboxyl	791	791	100%	0 0	98%
JF944700.1	Thladiantha nudiflora voucher LiHT080199 ribulose-1,5-bisphosphate carb	791	791	100%	0 0	98%
JF944699.1	Thladiantha nudiflora voucher LiHT1013 ribulose-1,5-bisphosphate carbo>	791	791	100%	0 0	98%
JF944698.1	Thladiantha nudiflora voucher LiHT1020 ribulose-1,5-bisphosphate carbo>	791	791	100%	0 0	98%
JF944697.1	Thladiantha nudiflora voucher LiHT1021 ribulose-1,5-bisphosphate carbo>	791	791	100%	0 0	98%
JF944696.1	Thladiantha nudiflora voucher LiHT1024 ribulose-1,5-bisphosphate carbo>	791	791	100%	0 0	98%
JF944694.1	Thladiantha nudiflora voucher LiHT1037 ribulose-1,5-bisphosphate carbo>	791	791	100%	0 0	98%
JF944689.1	Thladiantha montana voucher LiHT10162 ribulose-1,5-bisphosphate carb(	791	791	100%	0 0	98%
JF944688.1	Thladiantha montana voucher LiHT10163 ribulose-1,5-bisphosphate carb(	791	791	100%	0 0	98%
JF944687.1	Thladiantha montana voucher LiHT10166 ribulose-1,5-bisphosphate carb(	791	791	100%	0 0	98%
JF944686.1	Thladiantha montana voucher LiHT10167 ribulose-1,5-bisphosphate carb(	791	791	100%	0 0	98%
JF944685.1	Thladiantha maculata voucher LiHT298 ribulose-1,5-bisphosphate carbox	791	791	100%	0 0	98%
JF944684.1	Thladiantha maculata voucher LiHT305 ribulose-1,5-bisphosphate carbox	791	791	100%	0 0	98%
JF944683.1	Thladiantha maculata voucher LiHT306 ribulose-1,5-bisphosphate carbox	791	791	100%	0 0	98%
JF944682.1	Thladiantha maculata voucher LiHT307 ribulose-1,5-bisphosphate carbox	791	791	100%	0 0	98%
JF944681.1	Thladiantha maculata voucher LiHT314 ribulose-1,5-bisphosphate carbox	791	791	100%	0 0	98%
JF944680.1	Thladiantha maculata voucher LiHT315 ribulose-1,5-bisphosphate carbox	791	791	100%	0 0	98%
JF944676.1	Thladiantha lijiangensis voucher LiHT10176 ribulose-1,5-bisphosphate car	791	791	100%	0 0	98%
JF944675.1	Thladiantha lijiangensis voucher LiHT10178 ribulose-1,5-bisphosphate car	791	791	100%	0 0	98%
JF944674.1	Thladiantha lijiangensis voucher LiHT10174 ribulose-1,5-bisphosphate car	791	791	100%	0 0	98%
JF944673.1	Thladiantha lijiangensis voucher LiHT10175 ribulose-1,5-bisphosphate car	791	791	100%	0 0	98%

Figure 4. Blast analysis of the query sequence obtained by subjecting the amplified PCR products of bitter gourd which are sequenced by ABI377 sequencer.

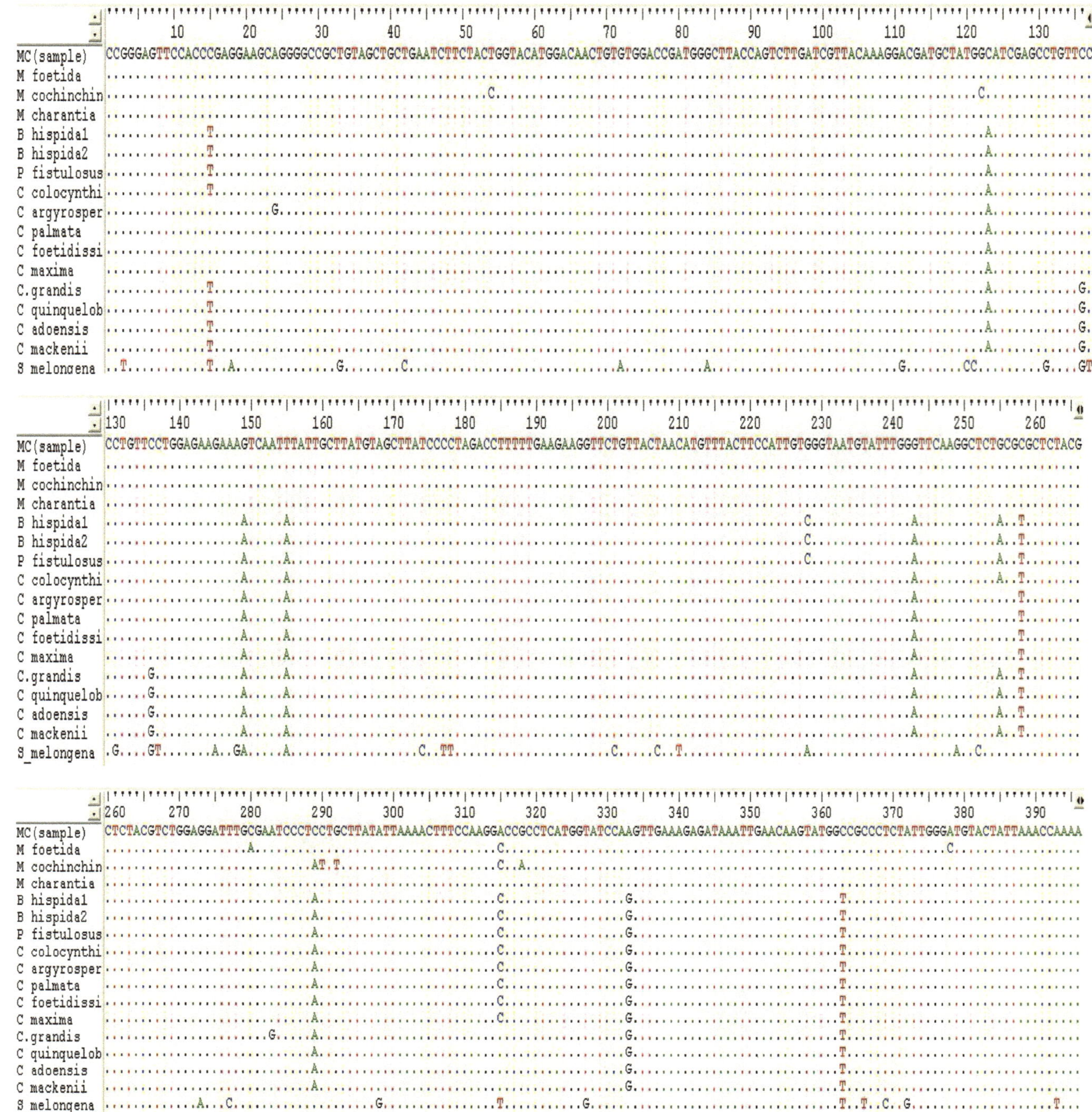

Figure 5. Multiple sequence alignment of the 17 input sequences using MEGA4 software with sequences bootstrapped to 390 base pairs (bitter gourd).

16

	1	2	3	4	5	6	7	8	9	10	11	12	13	14	15	16	17
1. MC																	
2. M foetida	0.007																
3. M cochinchinensis	0.016	0.018															
4. M charantia	0.000	0.007	0.016														
5. B hispida1	0.027	0.030	0.034	0.027													
6. B hispida2	0.027	0.030	0.034	0.027	0.000												
7. P fistulosus	0.027	0.030	0.034	0.027	0.000	0.000											
8. C colocynthis	0.025	0.027	0.032	0.025	0.002	0.002	0.002										
9. C argyrosperma	0.023	0.025	0.030	0.023	0.009	0.009	0.009	0.007									
10. C palmata	0.020	0.023	0.027	0.020	0.007	0.007	0.007	0.004	0.002								
11. C foetidissima	0.020	0.023	0.027	0.020	0.007	0.007	0.007	0.004	0.002	0.000							
12. C maxima	0.020	0.023	0.027	0.020	0.007	0.007	0.007	0.004	0.002	0.000	0.000						
13. C.grandis	0.027	0.034	0.039	0.027	0.009	0.009	0.009	0.007	0.014	0.011	0.011	0.011					
14. C quinqueloba	0.025	0.032	0.037	0.025	0.007	0.007	0.007	0.004	0.011	0.009	0.009	0.009	0.002				
15. C adoensis	0.025	0.032	0.037	0.025	0.007	0.007	0.007	0.004	0.011	0.009	0.009	0.009	0.002	0.000			
16. C mackenii	0.025	0.032	0.037	0.025	0.007	0.007	0.007	0.004	0.011	0.009	0.009	0.009	0.002	0.000	0.000		
17. S melongena	0.103	0.108	0.118	0.103	0.108	0.138	0.108	0.108	0.110	0.108	0.108	0.108	0.108	0.105	0.105	0.105	

Figure 6. Estimates of evolutionary divergences between 17 input sequences using maximum composite likelihood method in MEGA4 software (bitter gourd).

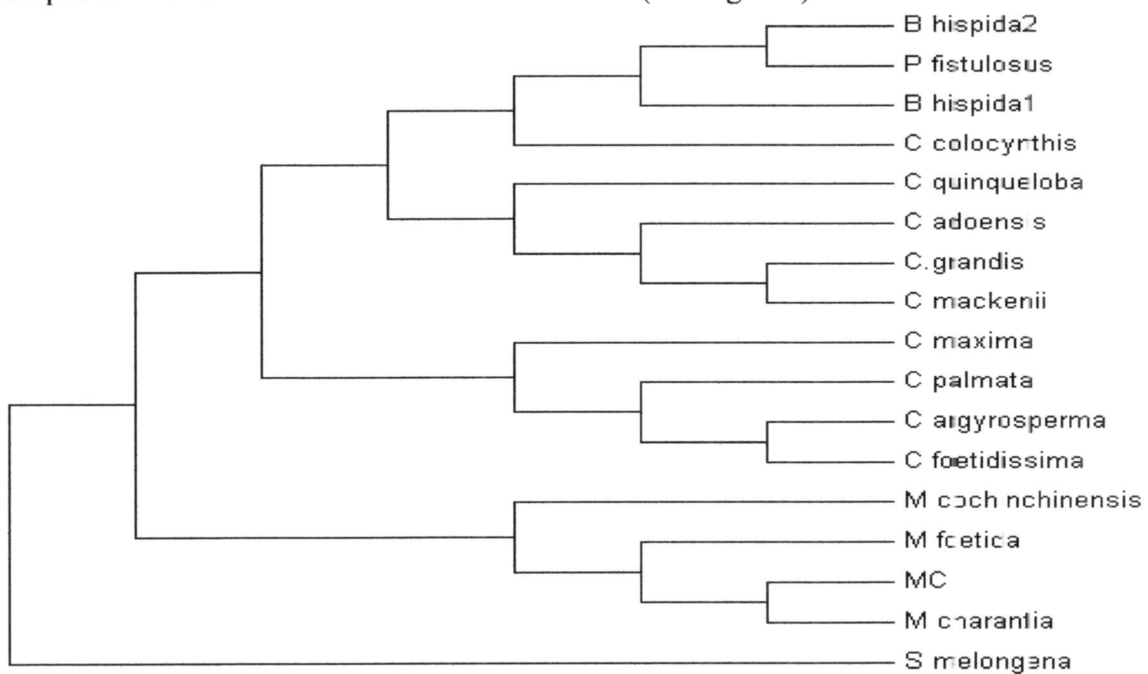

Figure 7. Evolutionary history of 17 taxa determined using neighbour-joining method (bitter gourd).

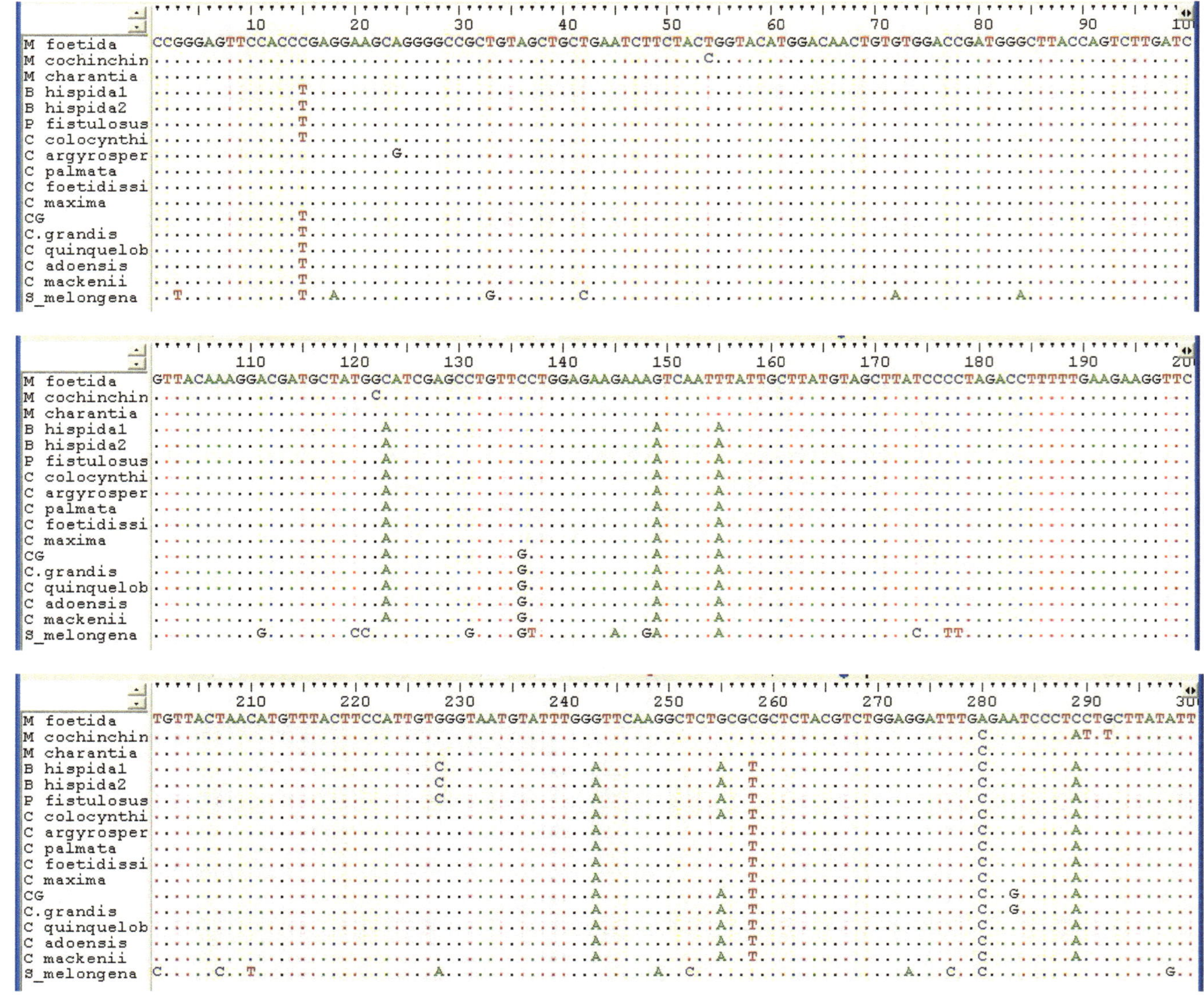

Figure 8. Multiple sequence alignment of the 16 input sequences and out-group using MEGA4 software with sequences bootstrapped to 300 base pairs (ivy gourd).

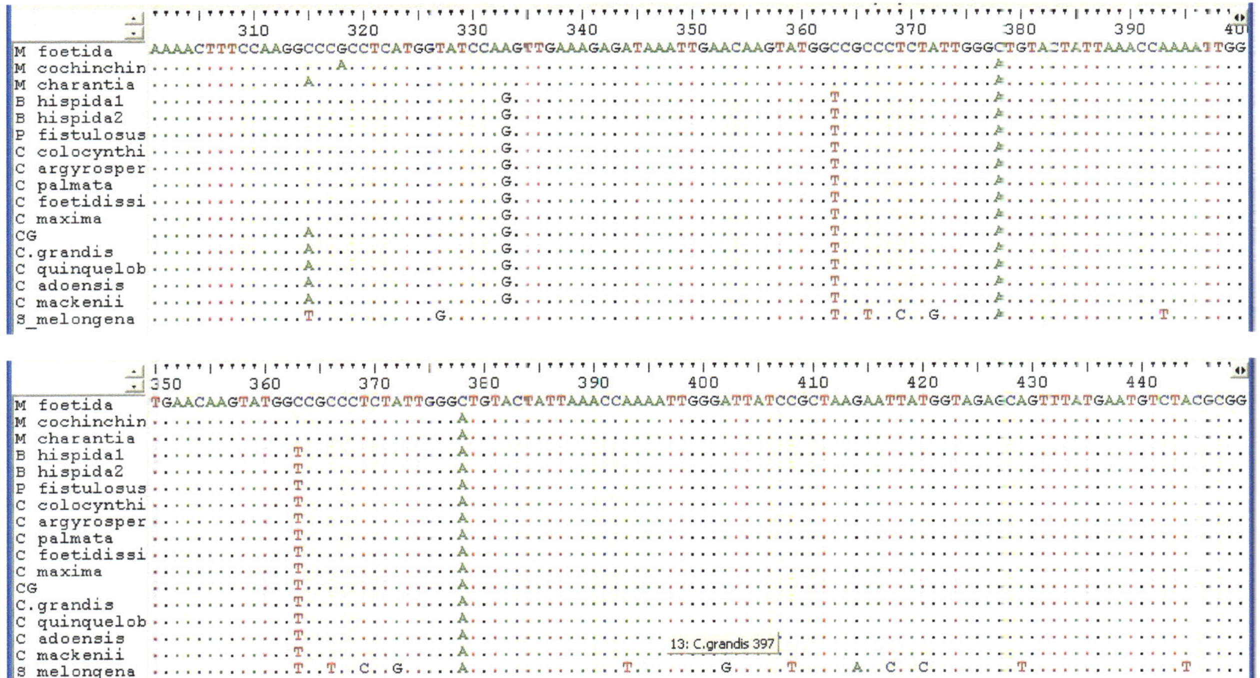

Figure 9. Multiple sequence alignment of the 16 input sequences and out-group using MEGA4 software with sequences bootstrapped to 440 base pairs (ivy gourd).

	1	2	3	4	5	6	7	8	9	10	11	12	13	14	15	16	17
1. M foetida																	
2. M cochinchinensis	0.018																
3. M charantia	0.007	0.016															
4. B hispida1	0.030	0.034	0.027														
5. B hispida2	0.030	0.034	0.027	0.000													
6. P fistulosus	0.030	0.034	0.027	0.000	0.000												
7. C colocynthis	0.027	0.032	0.025	0.002	0.002	0.002											
8. Sample	0.025	0.030	0.023	0.009	0.009	0.009	0.007										
9. C argyrosperma	0.025	0.030	0.023	0.009	0.009	0.009	0.007	0.000									
10. C palmata	0.023	0.027	0.020	0.007	0.007	0.007	0.004	0.002	0.002								
11. C foetidissima	0.023	0.027	0.020	0.007	0.007	0.007	0.004	0.002	0.002	0.000							
12. C maxima	0.023	0.027	0.020	0.007	0.007	0.007	0.004	0.002	0.002	0.000	0.000						
13. C.grandis	0.034	0.039	0.027	0.009	0.009	0.009	0.007	0.014	0.014	0.011	0.011	0.011					
14. C quinqueloba	0.032	0.037	0.025	0.007	0.007	0.007	0.004	0.011	0.011	0.009	0.009	0.009	0.002				
15. C adoensis	0.032	0.037	0.025	0.007	0.007	0.007	0.004	0.011	0.011	0.009	0.009	0.009	0.002	0.000			
16. C mackenii	0.032	0.037	0.025	0.007	0.007	0.007	0.004	0.011	0.011	0.009	0.009	0.009	0.002	0.000	0.000		
17. S melongena	0.108	0.119	0.103	0.108	0.108	0.108	0.108	0.110	0.110	0.108	0.108	0.108	0.108	0.105	0.105	0.105	

Figure 10. Estimates of evolutionary divergences between 16 input sequences and out-group using maximum composite likelihood method in MEGA4 software (ivy gourd).

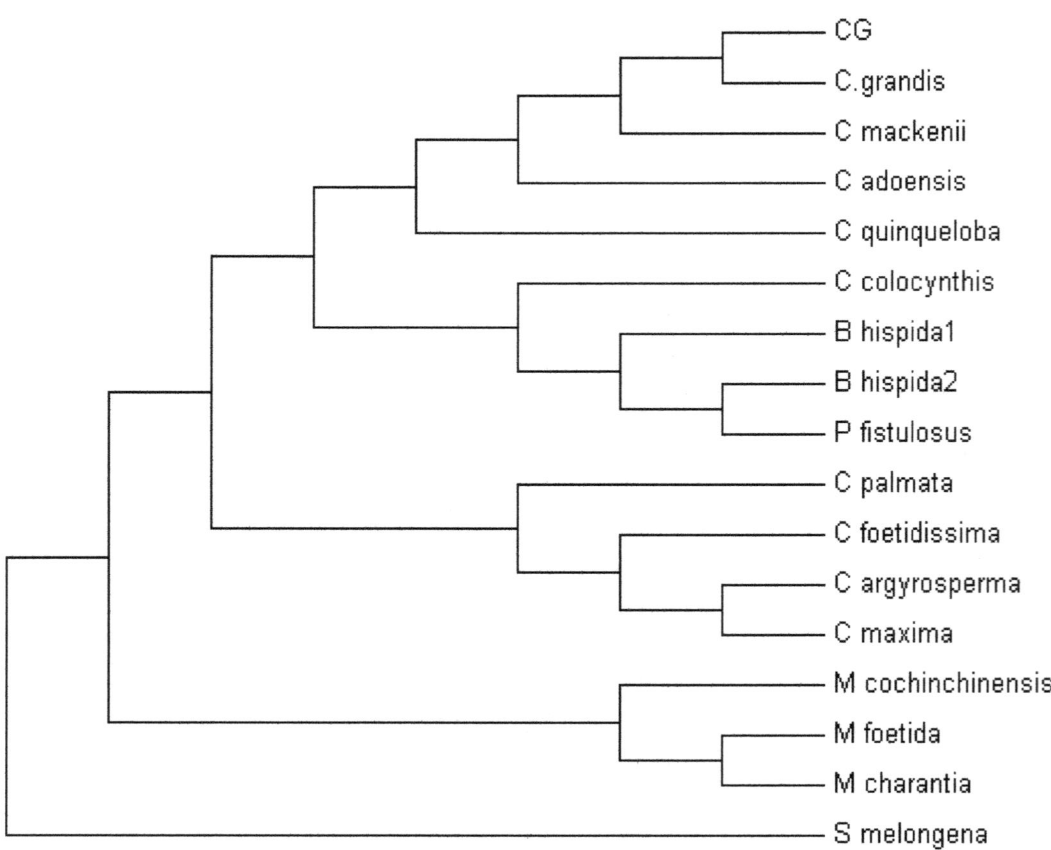

Figure 11. Evolutionary history of 17 taxa determined using neighbour-joining method (ivy gourd).

4. Discussion

Phylogenetic nomenclature is a result of Darwins discovery that history of life can be best represented in tree-shaped diagrams. In the current era, the conventional phyolgenetic nomenclatures are supplemented with molecular biological tools which make the classification much more precise and accurate. In general the DNA or RNA sequencing gained superiority among evolutionary studies coupled with the support of Bioinformatics tools and software's.

Currently, the sequencing of entire DNA of an organism is long and expensive process. A typical molecular phyolgenetic analysis require 1000 base pairs. At any location within such sequences, the bases found in a given position may vary between organisms. The particular sequences found in a given organism is referred as its halotype. In principle, since there are four base types, with 1000 base pair, there could have 41000 distinct halotypes. In a particular species or in a group of related species, there would have few variations and most of them are co-related with few distinct haplotypes.

Once divergences between all pairs of samples were determined, statistical cluster analysis and dendrograms examines the similarity among halotypes. Bootstrapping and jackknifing further increase the reliability estimates for the position of haplotypes within the evolutionary tree.

5. Conclusions

The reliability of molecular phylogenesis can be affected by myriad, long branch attraction, saturation and taxon sampling problems. So the selection of the models plays a key step in the success analysis. In general, the present study unambiguously confirmed the existence of *Coccinia grandis* as a distinct species and morphologic separation between *Coccinia grandis* and other selected members of Cucurbitaceae as separate species was evident. Among the species included in the present study *Coccinia grandis* undergone maximum divergence during the process of evolution. More detailed work on the evolution within the Cucurbitaceae is needed for crop improvement.

Acknowledgements

The authors are grateful for the cooperation of the management of Mar Augsthinose college for necessary support. Technical assistance from Binoy A Mulanthra is also acknowledged. We also thank an anonymous farmer for the supply of plant materials for our study.

References

Badola HK, Aitken S (2010). Biological resources and poverty alleviation in the Indian Himalayas. *Biodiversity* **11**(3-4) 8-18.

Baranek M, Stif, G, Vollmann J, Lelley T (2000). Genetic diversity within and between the species *Cucurbita pepo*, *C. moschata* and *C. maxima* as revealed by RAPD markers. *Report-Cucurbit Genetics Cooperative* **23**(1) 73-77.

Chenna R, Sugawara H, Koike T, Lopez R, Gibson TJ, Higgins DG, Thompson JD (2003). Multiple sequence alignment with the Clustal series of programs. Nucleic acids research, 31(13), 3497-3500.

Erlich HA (1989). *PCR technology. Principles and applications for DNA amplification.* Stockton press.

Esteras C, Nuez F, Picó B, YiHong W, Behera TK, Kole C (2011). Genetic diversity studies in Cucurbits using molecular tools. *Genetics, Genomics and Breeding of Cucurbits* **2**(1) 140-198.

Ferriol M, Pico MB, Nuez F (2003). Genetic diversity of some accessions of *Cucurbita maxima* from Spain using RAPD and SBAP markers.*Genetic Resources and Crop Evolution* **50**(3) 227-238.

Gong L, Paris HS, Nee MH, Stift G, Pachner M, Vollmann J, Lelley T (2012). Genetic relationships and evolution in *Cucurbita pepo* (pumpkin, squash, gourd) as revealed by simple sequence repeat polymorphisms. *Theoretical and Applied Genetics* **124**(5) 875-891.

Gong L, Paris HS, Stift G, Pachner M, Vollmann J, Lelley T (2013). Genetic relationships and evolution in Cucurbita as viewed with simple sequence repeat polymorphisms: the centrality of C. okeechobeensis. *Genetic Resources and Crop Evolution* **2**(1) 1-16.

Gwanama C, Labuschagn, MT, Botha AM (2000). Analysis of genetic variation in Cucurbita moschata by random amplified polymorphic DNA (RAPD) markers. *Euphytica* **113**(1) 19-24.

Haas SA, Hild M, Wright AP, Hain T, Talibi D, Vingron M (2003). Genome scale design of PCR primers and long oligomers for DNA microarrays. *Nucleic acids Research* **31**(19) 5576-5581.

Jeffrey C. (2005). A new system of Cucurbitaceae. *Bot. Zhurn, 90*, 332-335.

Kidwell KK, Osborn TC (1992). Simple plant DNA isolation procedures. In *Plant genomes: methods for genetic and physical mapping* (pp. 1-13). Springer Netherlands.

Kocyan A, Zhang LB, Schaefer H, Renner SS (2007). A multi-locus chloroplast phylogeny for the Cucurbitaceae and its implications for character evolution and classification. *Molecular Phylogenetics and Evolution* **44**(2) 553-577.

Lebeda A, Widrlechner MP, Staub J, Ezura H, Zalapa J, Krˇistkova E (2007). Cucurbits (Cucurbitaceae; Cucumis spp., Cucurbita spp., Citrullus spp.). In: Singh RJ (ed) Genetic resources, chromosome engineering, and crop improvement: vegetable crops. CRC Press, Boca Raton, pp 271–376

Meier A, Sander P, Bottger EC (1996). Protocol 4. Elimination of contaminating DNA within PCR reagents. PCR Essential Techniques, John Wiley & Sons, New York, 19.

Paris HS, Yonash N, Portnoy V, Mozes-Daube N, Tzuri G, Katzir N (2003). Assessment of genetic relationships in *Cucurbita pepo* (Cucurbitaceae) using DNA markers. *Theoretical and Applied Genetics* **106**(6) 971-978.

Porebski S, Bailey LG, Baum BR (1997). Modification of a CTAB DNA extraction protocol for plants containing high polysaccharide and polyphenol components. *Plant Molecular Biology Reporter* **15**(1) 8-15.

Robinson RW, Decker-Walters DS (1997). *Cucurbits.* Cab international. Wallingford.

Ruiz RAT, Hemleben V (1991). Use of ribosomal DNA spacer probes to distinguish cultivars of Cucurbita pepo L. and other Cucurbitaceae. *Euphytica* **53**(1) 11-17.

Rychlik WJSW, Spencer WJ, Rhoads RE (1990). Optimization of the annealing temperature for DNA amplification in vitro. *Nucleic acids Research* **18**(21) 6409-6412.

Saiki RK (1989). The design and optimization of the PCR. PCR technology: Principles and applications for DNA amplification, 7-16.

Thompson JD, Higgins DG, Gibson TJ (1994). CLUSTAL W: improving the sensitivity of progressive multiple sequence alignment through sequence weighting, position-specific gap penalties and weight matrix choice. *Nucleic acids Research* **22**(22) 4673-4680.

Wilson HD (1989). Discordant patterns of allozyme and morphological variation in Mexican Cucurbita. *Systematic Botany* **1**(2) 612-623.

Wu J, Chang Z, Wu Q, Zhan H, Xie S (2011). Molecular diversity of Chinese *Cucurbita moschata* germplasm collections detected by AFLP markers. *Scientia Horticulturae* **128**(1) 7-13.

Zhang LB, Renner S (2003). The deepest splits in Chloranthaceae as resolved by chloroplast sequences. *International Journal of Plant Sciences* **164**(S5) S383-S392.

YOUR KNOWLEDGE HAS VALUE

- We will publish your bachelor's and master's thesis, essays and papers

- Your own eBook and book - sold worldwide in all relevant shops

- Earn money with each sale

Upload your text at www.GRIN.com and publish for free